EXTENDED SUMMARY

SEVERE SPACE WEATHER EVENTS—
UNDERSTANDING SOCIETAL AND ECONOMIC IMPACTS

A WORKSHOP REPORT

Committee on the Societal and Economic Impacts of Severe Space Weather Events: A Workshop

Space Studies Board

Division on Engineering and Physical Sciences

NATIONAL RESEARCH COUNCIL
OF THE NATIONAL ACADEMIES

THE NATIONAL ACADEMIES PRESS
Washington, D.C.
www.nap.edu

THE NATIONAL ACADEMIES PRESS   500 Fifth Street, N.W.   Washington, DC 20001

This booklet is based on the report *Severe Space Weather Events—Understanding Societal and Economic Impacts: A Workshop Report* (The National Academies Press, Washington, D.C., 2008). Material in the booklet was derived from and largely reproduces material from the original report, with additions limited to factual details intended to provide illustrative background information. Neither document offers advice that should be construed as conclusions or recommendations of the National Research Council or of the agency that provided support for the project.

The project of which this booklet is a part was supported by Contract NNH06CE15B between the National Academy of Sciences and the National Aeronautics and Space Administration.

The Committee on the Societal and Economic Impacts of Severe Space Weather Events: A Workshop extends special thanks to member William S. Lewis, who prepared the text of the booklet, and to Estelle Miller of the National Academies Press, who did the design and layout.

Cover: (Upper left) A looping eruptive prominence blasted out from a powerful active region on July 29, 2005, and within an hour had broken away from the Sun. Active regions are areas of strong magnetic forces. Image courtesy of SOHO, a project of international cooperation between the European Space Agency and NASA.

International Standard Book Number 13:   978-0-309-13811-6
International Standard Book Number 10:   0-309-13811-6

Copies of this booklet are available free of charge from:

Space Studies Board
National Research Council
500 Fifth Street, N.W.
Washington, DC 20001

Additional copies of this booklet are available for purchase from the National Academies Press, 500 Fifth Street, N.W., Lockbox 285, Washington, DC 20055; (800) 624-6242 or (202) 334-3313 (in the Washington metropolitan area); Internet, http://www.nap.edu.

Copyright 2009 by the National Academy of Sciences. All rights reserved.

Printed in the United States of America

THE NATIONAL ACADEMIES
Advisers to the Nation on Science, Engineering, and Medicine

The **National Academy of Sciences** is a private, nonprofit, self-perpetuating society of distinguished scholars engaged in scientific and engineering research, dedicated to the furtherance of science and technology and to their use for the general welfare. Upon the authority of the charter granted to it by the Congress in 1863, the Academy has a mandate that requires it to advise the federal government on scientific and technical matters. Dr. Ralph J. Cicerone is president of the National Academy of Sciences.

The **National Academy of Engineering** was established in 1964, under the charter of the National Academy of Sciences, as a parallel organization of outstanding engineers. It is autonomous in its administration and in the selection of its members, sharing with the National Academy of Sciences the responsibility for advising the federal government. The National Academy of Engineering also sponsors engineering programs aimed at meeting national needs, encourages education and research, and recognizes the superior achievements of engineers. Dr. Charles M. Vest is president of the National Academy of Engineering.

The **Institute of Medicine** was established in 1970 by the National Academy of Sciences to secure the services of eminent members of appropriate professions in the examination of policy matters pertaining to the health of the public. The Institute acts under the responsibility given to the National Academy of Sciences by its congressional charter to be an adviser to the federal government and, upon its own initiative, to identify issues of medical care, research, and education. Dr. Harvey V. Fineberg is president of the Institute of Medicine.

The **National Research Council** was organized by the National Academy of Sciences in 1916 to associate the broad community of science and technology with the Academy's purposes of furthering knowledge and advising the federal government. Functioning in accordance with general policies determined by the Academy, the Council has become the principal operating agency of both the National Academy of Sciences and the National Academy of Engineering in providing services to the government, the public, and the scientific and engineering communities. The Council is administered jointly by both Academies and the Institute of Medicine. Dr. Ralph J. Cicerone and Dr. Charles M. Vest are chair and vice chair, respectively, of the National Research Council.

www.national-academies.org

**COMMITTEE ON THE SOCIETAL AND ECONOMIC IMPACTS OF
SEVERE SPACE WEATHER EVENTS: A WORKSHOP**

DANIEL N. BAKER, University of Colorado at Boulder, *Chair*
ROBERTA BALSTAD, Center for International Earth Science Information Network,
    Columbia University
J. MICHAEL BODEAU, Northrop Grumman Space Technology
EUGENE CAMERON, United Airlines, Inc.
JOSEPH F. FENNELL, Aerospace Corporation
GENENE M. FISHER, American Meteorological Society
KEVIN F. FORBES, Catholic University of America
PAUL M. KINTNER, Cornell University
LOUIS G. LEFFLER, North American Electric Reliability Council (retired)
WILLIAM S. LEWIS, Southwest Research Institute
JOSEPH B. REAGAN, Lockheed Missiles and Space Company, Inc. (retired)
ARTHUR A. SMALL III, Pennsylvania State University
THOMAS A. STANSELL, Stansell Consulting
LEONARD STRACHAN, JR., Smithsonian Astrophysical Observatory

Staff

SANDRA J. GRAHAM, Study Director
THERESA M. FISHER, Program Associate
VICTORIA SWISHER, Research Associate
CATHERINE A. GRUBER, Editor

SPACE STUDIES BOARD

CHARLES F. KENNEL, Scripps Institution of Oceanography,
　University of California, San Diego, *Chair*
A. THOMAS YOUNG, Lockheed Martin Corporation (retired), *Vice Chair*
DANIEL N. BAKER, University of Colorado at Boulder
STEVEN J. BATTEL, Battel Engineering
CHARLES L. BENNETT, Johns Hopkins University
YVONNE C. BRILL, Aerospace Consultant
ELIZABETH R. CANTWELL, Oak Ridge National Laboratory
ANDREW B. CHRISTENSEN, Dixie State College and Aerospace Corporation
ALAN DRESSLER, Observatories of the Carnegie Institution
JACK D. FELLOWS, University Corporation for Atmospheric Research
FIONA A. HARRISON, California Institute of Technology
JOAN JOHNSON-FREESE, Naval War College
KLAUS KEIL, University of Hawaii
MOLLY K. MACAULEY, Resources for the Future
BERRIEN MOORE III, Climate Central
ROBERT T. PAPPALARDO, Jet Propulsion Laboratory
JAMES PAWELCZYK, Pennsylvania State University
SOROOSH SOROOSHIAN, University of California, Irvine
JOAN VERNIKOS, Thirdage LLC
JOSEPH F. VEVERKA, Cornell University
WARREN M. WASHINGTON, National Center for Atmospheric Research
CHARLES E. WOODWARD, University of Minnesota
ELLEN G. ZWEIBEL, University of Wisconsin

RICHARD ROWBERG, Interim Director

Contents

The Societal Context	3
The Impact of Space Weather	4
Industry-Specific Space Weather Impacts, 4	
Electric Power Industry, 4	
Spacecraft Operations, 5	
Airline Operations, 6	
Space-Based Positioning, Navigation, and Timing, 8	
Future Vulnerabilities: The Specter of Extreme Space Weather Past, 9	
Collateral Impacts of Severe Space Weather, 11	
Space Weather Infrastructure	17
Space Weather Forecasting: Capabilities and Limitations, 19	
Space Weather Models, 19	
Understanding the Societal and Economic Impacts of Severe Space Weather	22
References	23

EXTENDED SUMMARY

SEVERE SPACE WEATHER EVENTS—
UNDERSTANDING SOCIETAL AND ECONOMIC IMPACTS

A WORKSHOP REPORT

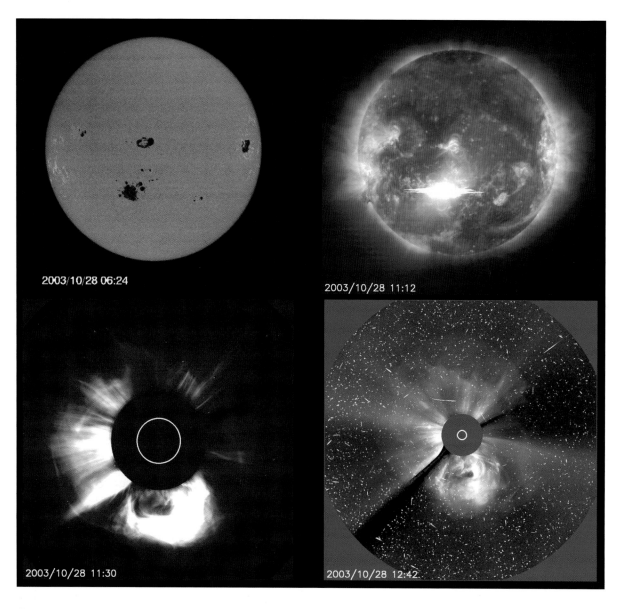

Genesis of a major space storm. On October 28, 2003, a large sunspot group in the Sun's southern hemisphere (upper left) erupted, producing an intense x-ray flare (upper right) and a large, fast "halo" coronal mass ejection (CME). Within less than an hour of CME lift-off/flare eruption, solar energetic particles (SEPs), accelerated by the shock wave preceding the CME, began arriving at Earth, causing a polar cap absorption event and initiating ozone-destroying chemistry in the middle atmosphere over the poles. The SEP event was still in progress the following day, when the CME slammed into Earth's magnetic field, triggering a powerful geomagnetic storm. The geomagnetic field was recovering when, late on October 30, a second CME, launched from the same active region as the first, arrived, unleashing another intense magnetic storm. All images were obtained with instruments on the Solar and Heliospheric Observatory (SOHO). Clockwise from the upper left: Michelson Doppler Imager (MDI) white light image of the photosphere; Extreme-ultraviolet Imaging Telescope (EIT) image of coronal Fe XII emissions; Large Angle and Spectrographic Corona (LASCO) C3 coronagraph image, showing "snow" caused by solar energetic particle bombardment; and a LASCO C2 near-Sun image of the halo CME. (Images courtesy of NASA/ESA.)

THE SOCIETAL CONTEXT

As evidenced in both ancient legend and the historical record, human activities, institutions, and technologies have always been prey to the extremes of weather—to droughts and floods, ice storms and blizzards, hurricanes and tornadoes. Around the middle of the 19th century, however, society in the developed parts of the world became vulnerable to a different kind of extreme weather as well—to severe disturbances of the upper atmosphere and the near-Earth space environment driven by the magnetic activity of the Sun. Although the nature of the solar-terrestrial connection was not understood at the time, such disturbances were quickly recognized as the culprit behind the widespread disruptions that periodically plagued the newly established and rapidly expanding telegraph networks. During the following century and a half, with the growth of the electric power industry, the development of telephone and radio communications, and a growing dependence on space-based communications and navigation systems, the vulnerability of modern society and its technological infrastructure to space weather has increased dramatically.

The effects of space weather on modern technological systems are well documented in both the technical literature and popular accounts. Most often cited perhaps is the collapse within 90 seconds of northeastern Canada's Hydro-Québec power grid during the great geomagnetic storm of March 1989, which left millions of people without electricity for up to 9 hours. This event exemplifies the dramatic impact that severe space weather can have on a technology upon which modern society in all of its manifold and interconnected activities and functions critically depends.

Nearly two decades have passed since the March 1989 event. During that time, awareness of the risks of severe space weather has increased among the affected industries, mitigation strategies have been developed, new sources of data have become available (e.g., the upstream solar wind measurements from the Advanced Composition Explorer), new models of the space environment have been created, and a national space weather infrastructure has evolved to provide data, alerts, and forecasts to an increasing number of users.

Now, 20 years later and approaching a new interval of increased solar activity, how well equipped are we to manage the effects of space weather? Have recent technological developments made our critical technologies more or less vulnerable? How well do we understand the broader societal and economic impacts of severe space weather events? Are our institutions prepared to cope with the effects of a "space weather Katrina," a rare, but according to the historical record, not inconceivable eventuality? On May 22 and 23, 2008, a one-and-a-half-day workshop held in Washington, D.C., under the auspices of the National Research Council's (NRC's) Space Studies Board brought together representatives of industry, the federal government, and the social science community to explore these and related questions. The key themes, ideas, and insights that emerged during the presentations and discussions are summarized in *Severe Space Weather Events—Understanding Societal and Economic Impacts: A Workshop Report* (The National Academies Press, Washington, D.C., 2008), which was prepared by the Committee on the Societal and Economic Impacts of Severe Space Weather Events: A Workshop. The present document is an expanded summary of that report.

THE IMPACT OF SPACE WEATHER

Modern technological society is characterized by a complex interweave of dependencies and interdependencies among its critical infrastructures. A complete picture of the socioeconomic impact of severe space weather must include both direct, industry-specific effects (such as power outages and spacecraft anomalies) and the collateral effects of space-weather-driven technology failures on dependent infrastructures and services.

Industry-Specific Space Weather Impacts

The electric power, spacecraft, and aviation industries are the main industries whose operations can be adversely affected by severe space weather. The effects of space weather can also be experienced by the growing number of users of the Global Positioning System (GPS) such as the oil and gas industry, which relies on GPS positioning data to support offshore drilling operations.

Electric Power Industry

During intense geomagnetic storms, the auroral oval moves to lower, more densely populated latitudes, where rapidly varying ionospheric currents associated with the aurora can produce direct-current flows in the electrical power grid. Such geomagnetically induced currents (GICs) can overload the grid, causing severe voltage regulation problems and, potentially, widespread power outages. Moreover, GICs can cause intense internal heating in extra-high-voltage (EHV) transformers, putting them at risk of failure or even permanent damage.

The March 1989 Quebec blackout referred to above remains the classic example of the impact of a severe space weather event—the most intense storm of the space age[1]—on the electric power industry. Storm-related GICs caused a voltage depression in the Hydro-Québec grid that Hydro-Québec's automatic voltage compensation equipment could not mitigate, resulting in a precipitous voltage collapse over a wide area. Specifically, five transmission lines from the James Bay hydroelectric power generation stations were tripped, causing a generation loss of 9,450 MW. With a load of about 21,350 MW, the system was unable to withstand the loss and collapsed within a minute and a half, blacking out the province of Quebec for approximately 9 hours. The effects of the storm were felt in the United States as well, in the Northeast, the upper-Midwest, the mid-Atlantic region, and even as far south as southern California. Approximately 200 storm-related events were reported to have affected power systems in North America; of these events the most severe was the failure of a large step-up transformer at the Salem Nuclear Power Plant in New Jersey. Other events ranged from generators tripping out of service, to voltage swings at major substations, to other, lesser equipment failures.

Following the 1989 collapse of the Hydro-Québec grid, electric power companies developed operational procedures to protect power grids against disruption and damage by severe space weather. Grid operators receive space weather forecasts from the Space Weather Prediction Center (SWPC) of the National Oceanic and Atmospheric Administration (NOAA) and from commercial and other space weather services. They also monitor voltages and ground currents in real time. During the geomagnetic storms of October and November 2003, for example, power grid operators in New England responded to severe space weather alerts and to real-time data from GIC monitors by modifying power grid operations in order to maintain adequate power quality for customers and reserve capacity to counteract the effects of the storms. Despite severe GICs, the power transmission equipment was protected, and the grid maintained continuous operation.

Spacecraft Operations

In late October 2003, powerful solar flares and fast Earthward-directed coronal mass ejections (CMEs) originating in an unusually large sunspot region (see p. 2) triggered especially intense geomagnetic and radiation storms during which more than half the spacecraft anomalies reported for that year occurred (Figure 1). The impact of space weather on spacecraft systems is not limited to dramatic CME-driven space weather events such as the 2003 "Halloween" storms and the March 1989 storm. Of major concern to the spacecraft industry are the periodic enhancements of the magnetospheric energetic electron environment associated with high-speed solar wind streams emanating from coronal holes during the declining phase of the solar cycle as well the injection of energetic plasma into the inner magnetosphere during magnetic substorms, which can occur during nonstorm times as well as storm times.

The effect of space weather on spacecraft operations is illustrated by the outage in January 1994 of two Canadian telecommunications satellites in geostationary orbit.[2] On January 20, 1994, Telesat's Anik E1 was disabled for about 7 hours as a result of damage to its control electronics by the discharge of electric charge deposited in the interior of the spacecraft by penetrating high-energy electrons. The outage occurred during an energetic electron storm that had begun a week earlier as a high-speed solar wind stream swept past Earth. During the E1 outage, the Canadian press was unable to deliver news to 100 newspapers and 450 radio stations. In addition, telephone service to 40 communities was interrupted. Shortly after E1 was restored to service, its sister satellite, Anik E2, went off the air, resulting in the loss of television and data services to more than 1,600 remote communities. Backup systems were also damaged, making the $290 million satellite useless. Approximately 100,000 home satellite dish owners were required to re-point their dishes manually to E1 and other satellites. It took Telesat operators 6 months to restore Anik E2 to service. The E2 failure is estimated to have cost Telesat $50 million to $70 million (U.S. dollars) in recovery costs and lost business.

The principal cause of space-weather-related spacecraft anomalies and failures is radiation in the form of solar energetic particles, galactic cosmic rays, and energetic particles trapped within Earth's radiation belts or accelerated during magnetospheric substorms. In order to design spacecraft that can withstand the effects of continuous exposure to space radiation and operate 24/7 for 10 to 15 years, spacecraft designers need accurate long-term models of the radiation environment and information about the statistical distribution of extreme events (e.g., the space weather equivalent of the "100-year storm"). Designers are thus concerned primarily with space climatology rather than with specific space weather events. Spacecraft operators, however, require real-time knowledge of the space environment as well as short-term forecasts ("nowcasts") in order to make operational decisions (e.g., with respect to thruster firing to reposition a spacecraft) that can reduce risks to spacecraft during disturbed conditions. (Such information is also used to support launch go/no-go decisions.) In the event of a spacecraft anomaly, knowledge of the

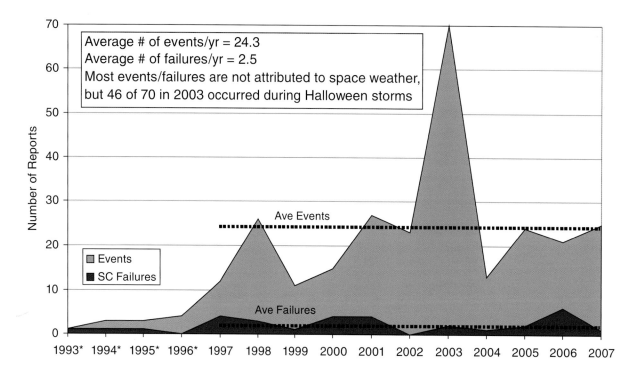

Figure 1. Telecommunication satellite anomalies and failures over a 14-year period. (Data for the years 1993-1996 are less extensive than for the period from 1997 on). The annual probability of an anomaly is around 10 percent, and the annual probability of a failure is about 1 percent. The big spike in 2003 reflects the anomalies that occurred during the October-November 2003 "Halloween" storms, which did not produce a significant rise in satellite failures. Around 250 commercial telecommunications satellites are operating in geosynchronous orbit. At a cost of roughly $300 million each, this fleet represents a $75 billion investment and generates an estimated annual revenue stream of more than $250 billion ($100 million per satellite per year). (Image courtesy of Michael Bodeau, Northrop Grumman.)

environment where the anomaly occurred as well as climatological information helps operators determine whether or not the anomaly was caused by space weather.

Airline Operations

In the late 1990s, airline companies began to fly polar routes between North America and Asia in order to avoid strong wintertime headwinds and thus to reduce travel time (Figure 2). Decreased travel time makes it possible to carry less fuel, thus saving costs, and allows the airlines to transport more passengers and cargo, increasing revenues. Because of the clear economic benefits, the use of polar routes has grown dramatically over the last decade. In 2007, thirteen carriers flew polar routes for a combined total of almost 7300 polar flights, an increase of nearly 2000 flights from the prior year.

The transpolar routes take aircraft to latitudes where satellite communication cannot be used, and flight crews must rely instead on high-frequency (HF) radio to maintain communication

Figure 2. Routes flown by transpolar flights between North America and Asia. Originally designated Polar 1, 2, 3, and 4, the routes were re-named after the waypoints ABERI, DEVID, RAMEL, and ORVIT. A fifth route, NIKIN (shown in red), was added in 2007. At latitudes above 82° (yellow circle), flight crews cannot use satellite communications and must rely instead on high-frequency (HF) radio to remain in contact with air traffic control. Changes in the polar ionosphere caused by solar energetic particle precipitation can degrade or totally black out HF radio communication. Transpolar flights must therefore be re-routed during intense solar radiation storms (solar energetic particle events). Timely space weather forecasts are important both for short-term (3-4 hour) operational planning and for longer-term (1 day) infrastructure planning (e.g., regarding air crew and aircraft assignments). (Image courtesy of Michael Stills, United Airlines.)

with the airline company and air traffic control, as required by federal regulation. During certain severe space weather events (referred to by the SWPC as "solar radiation storms"), solar energetic particles—primarily protons accelerated by CME-driven shocks—spiral down geomagnetic field lines into the polar ionosphere, where they increase the density of the ionized gas, which in turn affects the ability of the radio waves to propagate and can result in a complete radio blackout. Such polar cap absorption (PCA) events can last for several days, during which time aircraft must be diverted to latitudes where satellite communication links can be used. During several days of disturbed space weather in January 2005, for example, 26 United Airlines flights were diverted to nonpolar or less-than-optimum polar routes to avoid the risk of HF radio blackouts during PCA events. The increased flight time and extra landings and take-offs required by such route changes increased fuel consumption and raised cost, while the delays disrupted connections to other flights.

Space-Based Positioning, Navigation, and Timing

The 24 Global Positioning System satellites operated by the United States Air Force provide accurate positioning and timing information to a variety of military, government, and civilian users. In addition, "augmentations" by both commercial services and government agencies improve the accuracy, integrity, and availability of GPS data. For example, as part of the transition to space-based navigation as the primary means of navigation used by the National Airspace System, the Federal Aviation Administration (FAA) has implemented the Wide Area Augmentation System (WAAS), which provides precision horizontal and vertical navigation service over the continental United States, Alaska, and most of Canada and Mexico. WAAS effectively increases the capacity of the aviation system by allowing for reduced horizontal and vertical separation standards between planes without additional risk and by providing highly accurate vertical positioning that enables precision approaches and landings.

Current GPS-based navigation and positioning systems are vulnerable to space weather—specifically, to ionospheric density irregularities that affect the propagation of the signals from the GPS satellites to the receivers on the ground. Such irregularities are a routine occurrence near the equator; during magnetic storms, however, they occur in the midlatitude ionosphere as well. Degradation of the GPS signal by ionospheric irregularities produces ranging errors and can result in the temporary loss of GPS reception. Solar radio bursts have recently been found to be an additional source of interference with GPS reception in Earth's sunlit hemisphere.

Systems that use single-frequency receivers without augmentation are vulnerable even to minor ionospheric disturbances. Augmented systems are less susceptible to disruption by minor and moderate ionospheric disturbances but still can be adversely affected by scintillation, solar radio bursts, and major ionospheric disturbances. Thus, when WAAS detects ionospheric disturbances, it disables the use of precision navigation in the affected areas so that safety is never compromised. When large areas of disturbance are detected, precision navigation is disabled for all areas until 8 hours after disturbances cease. During the October 2003 magnetic storms, for example, WAAS vertical navigation service was disabled for approximately 30 hours, although horizontal navigation guidance was continuously available (Figures 3 and 4).

To mitigate the effects of space weather on the GPS, new signals and codes are being implemented that will allow GPS receivers to remove ionospheric ranging errors. This capability is expected to make augmentation systems unnecessary. In addition, the new signals and codes will be more resistant to fades caused by scintillation or solar radio bursts. The implementation of the new codes and signals, including the L5 signal dedicated to aviation, will take place incrementally over the next decade.

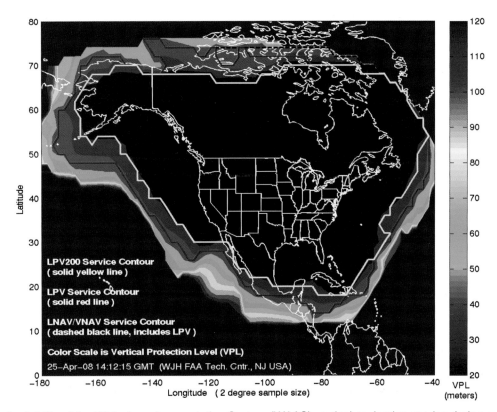

Figure 3. Availability of the Wide Area Augmentation System (WAAS) vertical navigation service during a geomagnetically quiet period. Vertical navigation for precision approaches (LPV) is available when the vertical protection level (VPL) is less than or equal to 50 meters; for WAAS-enabled approaches with a decision altitude down to 200 feet (LPV200) the VPL must be less than or equal to 35 meters. (LPV, localizer performance with vertical guidance; LNAV/VNAV, lateral and vertical navigation.) For LNAV/VNAV approaches the VPL must also be less than or equal to 50 meters. The horizonal protection level for LPV and LPV200 approaches—not shown—is 40 meters; for LNAV/VNAV it is 556 meters. (Image courtesy of Leo Eldredge, Federal Aviation Administration.)

In addition to its use in aviation, GPS positioning and timing information is widely used in a number of other applications, including precision farming, surveying and mapping, marine navigation, offshore drilling rig positioning, and transportation.

Future Vulnerabilities: The Specter of Extreme Space Weather Past

With increasing awareness and understanding of space weather and its effects on modern technological systems, vulnerable industries have adopted procedures and technologies designed to mitigate the impacts of space weather on their operations and customers. As noted above, airlines re-route flights scheduled for polar routes during intense solar energetic particle events in order to preserve reliable communications. Alerted to an impending geomagnetic storm and monitoring ground currents in real-time, power grid operators take defensive measures to protect the grid against GICs. Similarly, under adverse space weather conditions, launch personnel may

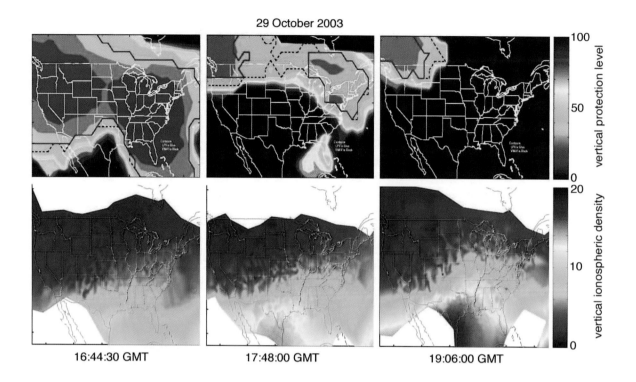

Figure 4. Progressive loss of vertical navigation service over North America (top row) as the ionospheric density disturbance (bottom row) worsens during the geomagnetic storm of October 29, 2003. Vertical navigation service over the continental U.S. was not fully restored until around 9:00 a.m. the following day. The color scale in the top panels shows the vertical protection level (VPL) measured in meters; the color scale in the bottom panels shows the vertical ionospheric density in meters. (Image adapted from material supplied by Leo Eldredge, Federal Aviation Adminstration.)

delay a launch, and satellite operators may postpone certain operations. For the spacecraft industry, however, the primary approach to mitigating space weather effects remains designing satellites to operate under extreme environmental conditions to the maximum extent possible within cost and resource constraints. GPS modernization through the addition of the new navigation signals and new codes will help mitigate space weather effects, although to what degree is not known. The FAA will therefore maintain "legacy" non-GPS-based navigation systems as a backup.

Our understanding of the vulnerabilities of modern technologies to severe space weather and the protective measures that have been developed are based largely on lessons learned during the past 20 or 30 years, during such episodes of severe space weather as the geomagnetic storms of March 1989 and October-November 2003. As severe as these recent events have been, the historical record reveals that space weather of even greater severity has occurred in the past (e.g., the "Carrington Event" of 1859 and the great magnetic storm of May 1921) and suggests that such extreme events, although rare, are likely to occur again some time in the future (see "The Great Magnetic Storms of August-September 1859," pp. 14-15).

It is not known how a severe space weather event far more intense than any experienced during the space age might impact our modern technological systems. Of particular concern is the degree to which the electric power grid, which lies at the heart of our national infrastructure, might be affected by such an event. A study by the Metatech Corporation suggests that, despite the protective procedures developed since the Hydro-Québec collapse, an unusually powerful magnetic storm could result in widespread outages and possible long-term damage to the nation's power grid. The Metatech study uses the great magnetic storm of May 1921 ("one of the greatest storms of the past ~130 years"[3]) to estimate the impact of an extreme space weather event on today's electric power grid. Using the rate of change in Earth's magnetic field measured in nanoteslas (nT) per minute as a proxy for GIC intensity, Metatech estimates that GICs during the 1921 storm would have been ten times more intense than those responsible for the March 1989 event. A storm of this magnitude today could result in large-scale blackouts affecting more than 130 million people (Figure 5). Moreover, according to the Metatech analysis, the intense GIC flows produced by the storm would place more than 300 large extra-high-voltage transformers at risk of failure or permanent damage, likely requiring a prolonged recovery period with long-term shortages of electric power to the affected areas (Figure 6).

Collateral Impacts of Severe Space Weather

An assessment of the societal and economic impacts of severe space weather must look beyond such direct space weather effects as spacecraft anomalies and power grid outages and consider how disruptions of vulnerable technological systems can affect the various sectors of society that are dependent on the functioning of these systems. Given the state of technology in the mid-19th century, the societal and economic impacts of the 1859 Carrington Event were limited to the disruptions of telegraph service "at the busy season when the telegraph is more than usually required,"[4] the telegraph companies' associated loss of income, and whatever the attendant effects on commerce might have been. Should an event of the magnitude of the Carrington Event occur today, the story could be quite different because of the central role that technology—in particular, electric power—plays in our society and because of the dependencies and interdependencies that characterize our critical infrastructures, rendering them vulnerable to failures cascading from one system to another.

Some of the indirect or collateral effects of a severe space weather event are vividly described in the following account of the 1989 Hydro-Québec blackout as it was experienced by the citizens of Montreal.

> The blackout closed schools and businesses, kept the Montreal Metro shut down during the morning rush hour, and paralyzed Dorval Airport, delaying flights. Without their navigation radar, no flight could land or take off until power had been restored. People ate their cold breakfast in the dark and left for work. They soon found themselves stuck in traffic that attempted to navigate darkened intersections without any streetlights or traffic control systems operating. . . . All these buildings [in downtown Montreal] were now pitch dark, stranding workers in offices, stairwells, and elevators. By some accounts, the blackout cost businesses tens of millions of dollars as it stalled production, idled workers, and spoiled products.[5]

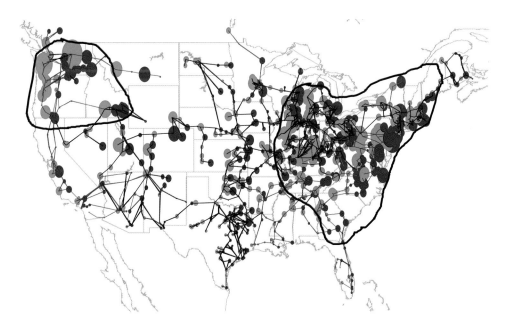

Figure 5. Regions susceptible to power grid collapse during a 4800 nT/min geomagnetic field disturbance at 50° geomagnetic latitude, where the densest part of the U.S. power grid lies. The affected regions are outlined in black. Analysis of such an event indicates that widespread blackouts could occur, involving more 130 million people. A disturbance of such magnitude, although rare, is not unprecedented: analysis of the May 1921 storm shows that disturbance levels of ~5000 nT/min were reached during that storm. (Image courtesy of John Kappenman, Metatech Corporation.)

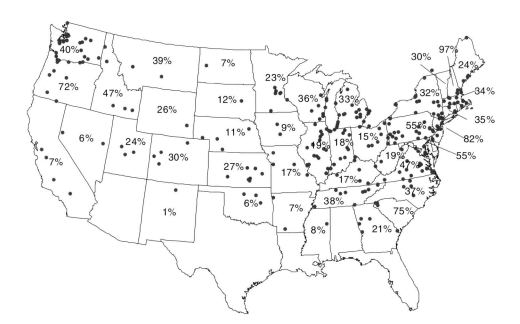

Figure 6. A map showing the extra-high-voltage transformer capacity (estimated at ~365 large transformers), by state, at risk of damage during a 4800 nT/min disturbance. Regions with high percentages could experience long-duration power outages lasting several years. (Image courtesy of John Kappenman, Metatech Corporation.)

A major power blackout, whether the result of severe space weather or severe terrestrial weather, has the potential to affect virtually all sectors of society: communications, transportation, banking and finance, commerce, manufacturing, energy, government, education, health care, public safety, emergency services, the food and water supply, and sanitation (Figure 7). The severity of the impacts depends on a number of variables, including the duration of the outage. The socioeconomic impacts of a long-term outage, requiring replacement of permanently damaged transformers, could be extensive and serious. According to an estimate by the Metatech Corporation, the total cost of a long-term, wide-area blackout caused by an extreme space weather event could be as much as $1 trillion to $2 trillion during the first year, with full recovery requiring 4 to 10 years depending on the extent of the damage. (For comparison, the total cost for the United States of the August 2003 blackout—a major non-space-weather-related blackout that affected 50 million people in the northeastern United States and Ontario—is estimated to have been between $4 billion and $10 billion.[6])

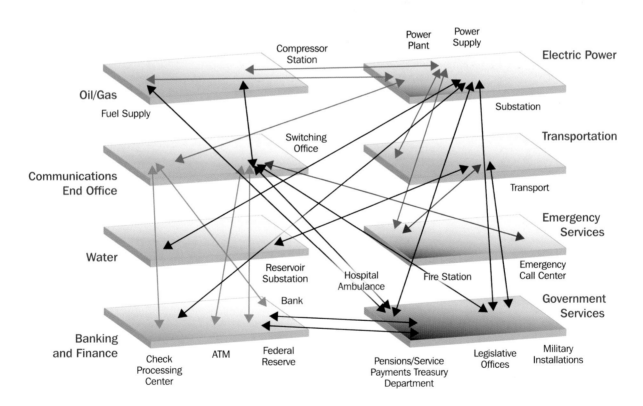

Figure 7. Schematic illustrating the interconnection of critical infrastructures and their dependencies and interdependencies. As the nation's infrastructures and services increase in complexity and interdependence over time, a major outage of any one infrastructure will have an increasingly widespread impact. (Image courtesy of Department of Homeland Security.)

The Great Magnetic Storms of August-September 1859 (the Carrington Event)[7]

Shortly after midnight on September 2, 1859, campers in the Rocky Mountains were awakened by an "auroral light, so bright that one could easily read common print." The campers' account, published in the *Rocky Mountain News,* continues, "Some of the party insisted that it was daylight and began the preparation of breakfast." Eighteen hundred miles to the east, Henry C. Perkins, a respected physician in Newburyport, Massachusetts, observed "a perfect dome of alternate red and green streamers" over New England. To the citizens of Havana, Cuba, the sky that night "appeared stained with blood and in a state of general conflagration." Dramatic auroral displays had been seen five nights before as well, on the night of August 28/29, when (again in the words of Dr. Perkins) "the whole celestial vault was glowing with streamers, crimson, yellow, and white, gathered into waving brilliant folds." In New York City, thousands gathered on sidewalks and rooftops to watch "the heavens . . . arrayed in a drapery more gorgeous than they have been for years." The aurora that New Yorkers witnessed that Sunday night, the *New York Times* assured its readers, "will be referred to hereafter among the events which occur but once or twice in a lifetime."[8]

Low-latitude red auroras, such as those widely reported to have been observed during the Carrington Event, are a characteristic feature of major geomagnetic storms. The aurora shown here was photographed over Napa Valley, California, during the magnetic storm of November 5, 2001. (Image courtesy D. Obudzinski, © Dirk Obudzinski 2001, www.borealis2000.com.)

From August 28 through September 4, auroral displays of extraordinary brilliance were observed throughout North and South America, Europe, Asia, and Australia, and were seen as far south as Hawaii, the Caribbean, and Central America in the Northern Hemisphere and in the Southern Hemisphere as far north as Santiago, Chile. Even after daybreak, when the aurora was no longer visible, its presence continued to be felt through the effect of the auroral currents. Magnetic observatories recorded disturbances in Earth's field so extreme that magnetometer traces were driven off scale, and telegraph networks around the world—the "Victorian Internet"[9]—experienced major disruptions and outages. "The electricity which attended this beautiful phenomenon took possession of the magnetic wires throughout the country," the *Philadelphia Evening Bulletin* reported, "and there were numerous side displays in the telegraph offices where fantastical and unreadable messages came through the instruments, and where the atmospheric fireworks assumed shape and substance in brilliant sparks."[10] In several locations, operators disconnected their systems from the batteries and sent messages using only the current induced by the aurora.[11]

The auroras were the visible manifestation of two powerful magnetic storms that occurred near the peak of the sunspot cycle. The two storms, which occurred in rapid succession, are referred as the "Carrington Event" in honor of Richard Carrington, a British amateur astronomer. On September 1, the day before the onset of the second storm, Carrington had observed an outburst of "two patches of intensely bright and white light"[12] from a large and complex group of sunspots near the center of the Sun's disk. Although the connection was not understood at the time, Carrington's observation provided the first evidence that eruptive activity on the Sun is the ultimate cause of geomagnetic storms.

We know today that what Carrington observed was an extraordinarily intense white-light flare that was associated with a powerful, fast-moving coronal mass ejection (CME). The CME and the shock wave that preceded it impacted Earth's magnetosphere some 17.5 hours after Carrington's observation, triggering an unusually severe geomagnetic storm. In addition to the low-latitude auroras and intense auroral currents responsible for the telegraph outages, all of the phenomena known today to be characteristic of a major magnetic storm occurred as well, although the mid-19th century lacked the means to detect and measure them, and its most sophisticated technologies were unaffected by them: an increased Earthward flow of magnetospheric plasma, creating or intensifying the ring current; the explosive release of stored magnetic energy in multiple magnetospheric substorms; an increase in the energy content of the radiation belts as well as the possible creation of temporary new belts; and changes in the ionospheric and thermospheric density at midlatitudes. Recent analysis of ice core data indicates that the geomagnetic storm was also accompanied by a solar energetic particle event four times more intense than the most severe solar energetic particle event of the space age. By this as well as other measures, the Carrington Event ranks as one of the most severe space weather events—and by some measures *the* most severe—on record.[13]

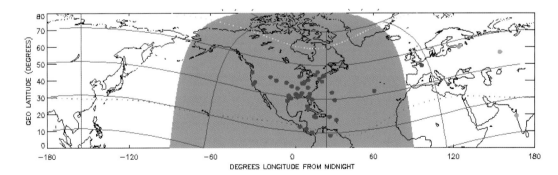

Locations of reported auroral observations during the first ~1.5 hours of the September 2, 1859, magnetic storm (orange dots). (Image courtesy of J.L. Green, NASA.)

> **Box 1**
> **Space Weather: Some Institutional Issues**
>
> Space weather potentially affects large complex technical systems that are vital for economic and social stability and functioning. But managing the effects of severe space weather is not just a technical problem: it is also, importantly, a problem of institutions and of society.
>
> A key issue affecting our ability to prevent disruption of large technical systems is the difficulty of developing the appropriate institutions to deal with the problem on a long-term basis. Institutional development occurs most often under conditions of frequent accidents or errors. When nothing bad appears to happen from one year to another, sustaining preparedness and planning in out-years is extraordinarily challenging. Consequently, space weather is not on the radar screen of many people outside the small technical community and some affected businesses.
>
> Dependency creep, risk migration, and new technologies are potential problems for operators of large technical systems. As systems become more complex, and as they grow in size, understanding and oversight become more difficult. Subsystems and dependencies may evolve that escape the close scrutiny of organization operators. Dependencies allow risk present in one part the other overall system to "migrate" to others, with potentially damaging results. GPS and electric power systems have clearly accelerated dependency creep, and consequent risk migration. New technologies, such as nanoscale components, may not be adequately understood in the context of 11-year solar cycles.
>
> One of the most fundamental concerns for operators of large technical systems is the efficiency-vulnerability tradeoff—that is, the question of how much reserve capacity is available to deal with uncertainty and contingencies. In stable protected environments, systems operate with excess capacity: costs are passed on to users and the society. In competitive-market but benign environments, however, systems operate at close to their efficiency frontiers. Slack resources are consumed, buffers shrink, costs fall, and profits rise. But in competitive-market and "hostile" environments where unexpected developments perturb the system, finely tuned technical systems become brittle and have trouble operating outside relatively narrow parameters. Vulnerability can be the consequence of increased efficiency. "Security externalities" emerge due to interdependencies, lack of knowledge, lack of slack, lack of trust, and lack of ways to overcome coordination problems.

Space storms of the magnitude of the Carrington Event are fortunately very rare, and the risk that such an event might cause a long-term catastrophic power grid collapse with major socioeconomic disruptions, while real, is low. In the field of risk analysis, such an extreme event is termed a low-frequency/high-consequence (LF/HC) event. In terms of their potential broader, collateral impacts, LF/HC events present a unique set of problems for public (and private) institutions and governance, different from the problems raised by conventional, expected, and frequently experienced events. As a consequence, dealing with the collateral impacts of LF/HC events requires different types of budgeting and management capabilities and consequently challenges the basis for conventional policies and risk management strategies, which assume a universe of constant or reliable conditions. Moreover, because systems can quickly become dependent on new technologies in ways that are unknown and unexpected by both developers and users, vulnerabilities in one part of the broader system have a tendency to spread to other parts of the system. Consequently, it is difficult to understand, much less to predict, the consequences of future LF/HC events. Sustaining preparedness and planning for such events in future years is equally difficult (Box 1).

SPACE WEATHER INFRASTRUCTURE

Space weather services in the United States are provided primarily by NOAA's SWPC and the U.S. Air Force's (USAF's) Weather Agency (AFWA), which work in close partnership to address the needs of their civilian and military user communities, respectively.

The SWPC draws on a variety of data sources, both space- and ground-based, to provide forecasts, watches, warnings, alerts, and summaries as well as operational space weather products to civilian and commercial users. Its primary sources of information about solar activity, upstream solar wind conditions, and the geospace environment are NASA's Advanced Composition Explorer (ACE), NOAA's GOES and POES satellites, magnetometers, and the USAF's solar observing networks. Secondary sources include SOHO and STEREO as well as a number of ground-based facilities. Despite a small and unstable budget (roughly $5 million to $6 million U.S. dollars annually) that limits capabilities, the SWPC has experienced a steady growth in customer base, even during the solar minimum years when disturbance activity is lower (Figure 8).

The focus of the USAF's space weather effort is on providing situational knowledge of the real-time space weather environment and assessments of the impacts of space weather on different Department of Defense (DOD) missions. The Air Force uses NOAA data combined with data from its own assets such as the Defense Meteorological Satellites Program (DMSP), the

Impact Area	Customer (examples)	Action (examples)	Cost (examples)
Spacecraft (Individual systems to complete spacecraft failure; communications and radiation effects)	• Lockheed Martin • Orbital • Boeing • Space Systems Loral • NASA, DoD	• Postpone launch • In orbit - Reboot systems • Turn off/safe instruments and/or spacecraft	• Loss of spacecraft ~$500M • Commercial loss exceeds $1B • Worst case storm - $100B
Electric Power (Equipment damage to electrical grid failure and blackout conditions)	• U.S. Nuclear Regulatory Commission • N. America Electric Reliability Corp. • Allegheny Power • New York Power Authority	• Adjust/reduce system load • Disconnect components • Postpone maintenance	• Estimated loss ~$400M from unexpected geomagnetic storms • $3-6B loss in GDP (blackout)
Airlines (Communications) (Loss of flight HF radio communications) (Radiation dose to crew and passengers)	• United Airlines • Lufthansa • Continental Airlines • Korean Airlines • NavCanada (Air Traffic Control)	• Divert polar flights • Change flight plans • Change altitude • Select alternate communications	• Cost ~ $100k per diverted flight • $10-50k for re-routes • Health risks
Surveying and Navigation (Use of magnetic field or GPS could be impacted)	• FAA-WAAS • Dept. of Transportation • BP Alaska and Schlumberger	• Postpone activities • Redo survey • Use backup systems	• From $50k to $1M daily for single company

Figure 8. Examples of impact areas and customers for space weather data provided by the Space Weather Prediction Center (SWPC). More than 6500 unique customers subscribe to the SWPC's product subscription service. Many data files and products are also available on an anonymous FTP server. Selected products are also distributed on the NOAA/National Weather Service dedicated broadcast systems. More than 50 million files are transferred from the SWPC web page each month. More than 500,000 files are created monthly with near-real-time data for 176 different products serving more than 400,000 unique customers every month in more than 120 countries. (Image courtesy of William Murtagh, NOAA Space Weather Prediction Center.)

Communications/Navigation Outage Forecasting System, the Solar Electro-Optical Network, the Digital Ionospheric Sounding System, and the GPS network (Figure 9).

NASA is the third major element in the nation's space weather infrastructure. Although NASA's role is scientific rather than operational, NASA science missions such as ACE provide critical space weather information, and NASA's Living with a Star program targets research and technologies that are relevant to operations. NASA-developed products that are candidates for eventual transfer from research to operations include physics-based space weather models that can be transitioned into operational tools for forecasting and situational awareness and sensor technology.

NOAA, NASA, and the Air Force are all involved in the National Polar Orbiting Environmental Satellite System (NPOESS), the joint civilian-military successor to the DMSP. Among its other objectives, NPOESS was intended to take over the DMSP's space weather monitoring function. However, in 2006, because of large and increasing cost overruns, the NPOESS program underwent a dramatic restructuring. (Even before these changes, compromises had been made with regard to some of the desired space environmental measurements.) The restructuring eliminated sensors and reduced the size of the on-orbit constellation from three spacecraft to two, resulting in a system that will have less capability to make critical measurements of the space environment than is currently available. The system is planned to last through 2024-2026, with the first NPOESS spacecraft to be launched 2013.

Space-Based Measurement

1 DMSP/SES*
2 ACE/SOHO FO
3 GOES
4 GPS
5 DSP
6 NPOESS
7 C/NOFS

Space Weather Parameter	Example Mission Supported	Observing Capability (Threshold SSA)	Forecasting Capability (Objective SSA)
Ionospheric Electrons (60%) 1, 2, 7	Geolocation		
Ionospheric Disturbances (60%) 1, 2, 7	Communications		
Energetic Particles (90%) 1, 2, 3, 4, 5, 6, 7	Satellite Operations		
Radiation & Disturbances (75%) 1, 2, 3, 4, 5, 6, 7	Space Tracking		
Ionospheric Disturbances (60%) 1, 2, 7	Navigation		

■ Good (>75%)   ■ Moderate (50-75%)   □ Marginal (25-50%)   ■ Little or None (0-25%)

Ground-Based Measurement

1 SOON/ISOON
2 RSTN/RSTN II
3 NEXION
4 TEC
5 SCINDA
6 Geomag

Space Weather Parameter	Example Mission Supported	Observing Capability (Threshold SSA)	Forecasting Capability (Objective SSA)
Ionospheric Electrons (60%) 1, 2, 3, 4, 5, 6	Geolocation		
Ionospheric Disturbances (60%) 1, 2, 3, 4, 5, 6	Communications		
Energetic Particles (25%) 1, 2, 6	Satellite Operations		
Radiation & Disturbances (40%) 1, 2, 3, 4, 5, 6	Space Tracking		
Ionospheric Disturbances (50%) 1, 2, 3, 4, 5, 6	Navigation		

*SES – Space Environment Sensors as payload on other satellites

Figure 9. Sources and types of space weather data needed to support representative military mission areas. Color coding indicates the Air Force Weather Agency's current capability level. (Image courtesy of Herbert Keyser, U.S. Air Force.)

In addition to NASA, NOAA, and the DOD, several other federal agencies (e.g., the National Science Foundation, the Department of Energy) are involved to varying degrees in the nation's space weather effort, which is coordinated through the National Space Weather Program under the auspices of the Office of the Federal Coordinator for Meteorology. Other key elements of the nation's space weather infrastructure are the solar and space physics research community and the emerging commercial space weather businesses. Of particular importance are the efforts of these sectors in the area of model development.

Space Weather Forecasting: Capabilities and Limitations

One of the important functions of a nation's space weather infrastructure is to provide reliable long-term forecasts, although the importance of forecasts varies according to industry. With long-term (1- to 3-day) forecasts and minimal false alarms, the various user communities can take actions to mitigate the effects of impending solar disturbances and to minimize their economic impact. Currently, NOAA's SWPC can make probability forecasts of space weather events with varying degrees of success (Figure 10, *top*). For example, the SWPC can, with moderate confidence, predict 1 to 3 days in advance the probability of occurrence of a geomagnetic storm or an X-class flare, whereas its capability to provide even short-term (less than 1 day) or long-term forecasts of ionospheric disturbances—information important for GPS users—is poor. The SWPC has identified a number of critical steps needed to improve its forecasting capability, enabling it, for example, to provide high-confidence long-term and short-term forecasts of geomagnetic storms and ionospheric disturbances (Figure 10, *bottom*). These steps include securing an operational solar wind monitor at L1; transitioning research models (e.g., of coronal mass ejection propagation, the geospace radiation environment, and the coupled magnetosphere-ionosphere-atmosphere system) into operations; and developing precision GPS forecast and correction tools (Box 2).

The requirement for a solar wind monitor at L1 is particularly important because ACE, the SWPC's sole source of real-time solar wind and interplanetary magnetic field data, is well beyond its planned operational life, and provisions to replace it have not been made. Positioned 1.5 million kilometers upstream from Earth, ACE provides a critical ~45 minutes of advanced warning before a CME strikes Earth. Recognizing the importance of an upstream monitor, Congress mandated in the 2008 NASA Authorization Act that the Office of Science and Technology Policy "develop a plan for sustaining space-based measurements of solar wind from the L-1 Lagrangian point in space and for the dissemination of the data for operational purposes." The plan is to be developed in consultation "with NASA, NOAA, and other Federal agencies, and with industry."

Although the SWPC does not classify SOHO as a primary data source, it relies heavily on SOHO coronographic observations to predict the properties and trajectories of CMEs responsible for large geomagnetic storms.

Space Weather Models

Successfully forecasting space weather requires the development of a suite of models covering the various domains of the space environment, from the solar corona to Earth's ionosphere and thermosphere. An area of particular interest is the implementation for operational use of

Long-Term Forecast (1-3 days)	Short-Term Forecasts and Warnings (<1 day)	Now-casts and Alerts
M-flare and X-flare probabilities	M-flare and X-flare probabilities	X-ray flux – global and regional
Solar energetic particle probabilities	Solar energetic particle probabilities	Energetic Particle Environment (protons and electrons) – global and regional
Geomagnetic storm probabilities	Geomagnetic storm probabilities – global and regional	Geomagnetic activity – global and regional
Ionospheric disturbance probabilities	Ionospheric disturbance probabilities – global and regional	Ionospheric disturbances (TEC, irregularities, HF propagation) – global and regional
Solar irradiance flux levels (EUV and 10.7 cm) (1-7 days for f10.7)		Solar irradiance (EUV and f10.7) – global

Long-Term Forecast (1-3 days)	Short-Term Forecasts and Warnings (<1 day)	Now-casts and Alerts
M-flare and X-flare probabilities	M-flare and X-flare	X-ray flux – global and regional
Solar energetic particle probabilities	Solar energetic particles	Energetic Particle Environment (protons and electrons) – global and regional
Geomagnetic storm probabilities	Geomagnetic storms – global and regional	Geomagnetic activity – global and regional
Ionospheric disturbance probabilities	Ionospheric disturbances – global and regional	Ionospheric disturbances (TEC, irregularities, HF propagation) – global and regional
Solar irradiance flux levels (EUV and 10.7 cm) (1-7 days for f10.7)		Solar irradiance (EUV and f10.7) – global

Figure 10. *Top:* Capability levels of NOAA's Space Weather Prediction Center in FY 2008. *Bottom:* Potential capability levels in FY 2014, assuming adequate funding to support the developments listed in Box 2. Green, satisfactory; yellow, less than satisfactory; red, poor. (Image courtesy of Thomas J. Bogdan, Space Weather Prediction Center, NOAA.)

> **Box 2**
> **Future Developments Identified by the Space Weather Prediction Center as Needed to Improve Its Forecasting and Prediction Capability**
>
> - Secure an operational L1 solar wind monitor.
> - Transition a numerical coronal mass ejection/solar wind model into operations.
> - Secure backup capability for GOES-10 XRS (X Ray Spectrometer) data stream.
> - Complete compliance measures necessary for the Space Weather Prediction Center to become a partner in the National Climate Service to help guide future solar observations, research, modeling, and forecast development activities.
> - Transition the whole-atmosphere model into operations.
> - Develop forecast capabilities based on STEREO data streams.
> - Revamp the concept of operations of the Space Weather Forecast Office.
> - Transition a coupled magnetosphere/whole-atmosphere model into operations.
> - Develop precision Global Positioning System forecast and correction tools.
> - Develop operational radiation environment models.

physics-based models, which can provide more accurate, longer-lead-time predictions of severe space weather conditions. In addition to physics-based forecast models, there is also a need for improved climatological models of Earth's radiation environment (Figure 11). As noted earlier (p. 5), radiation belt climatology models are of special importance to the spacecraft industry.

Models currently in use at the SWPC include the U.S. Total Electron Content model, which estimates the delays in GPS signals due to the changes in the electron content of the ionospheric path between the GPS satellite and the receiver, and the Wang-Sheeley-Arge model, which predicts solar wind speed and the polarity of the interplanetary magnetic field at Earth. These are two important quantities for determining the severity of geomagnetic disturbances caused by solar wind and CME events. Among the models implemented by AFWA are the Global Assimilation of Ionospheric Measurements (GAIM) model, which assimilates data from a variety of space- and ground-based sources to specify the ionospheric environment, and the Hakamada-Akasofu-Fry (HAF) solar wind model. The Space Weather Modeling System currently under development by the DOD will couple the HAF model with the physics-based ionospheric model within the GAIM model, enabling multiday forecasts of the ionospheric environment and its response to solar wind forcing.

To facilitate the transition of physics-based research models to operations, the multiagency Community Coordinated Modeling Center (CCMC) was established in the late 1990s. The CCMC tests and validates advanced space weather models developed by the research community and evaluates their usefulness for operations. The models hosted by the CCMC are available for use by the wider solar and space physics community.

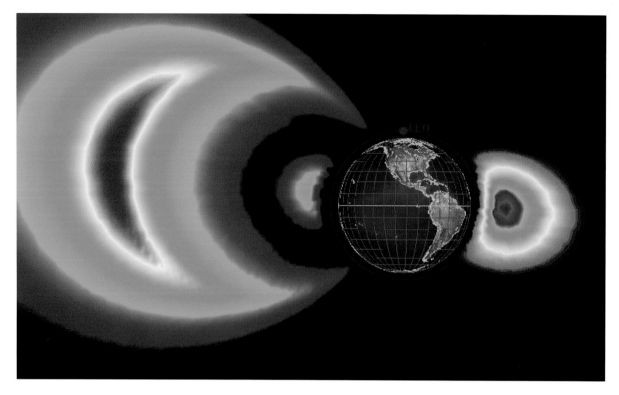

Figure 11. Earth's electron belt (left) and proton belt (right) as defined by the empirical radiation belt models available from NASA for use by satellite designers and mission planners. These models are based on older and limited data and do not incorporate data from more recent missions such as CRRES, SAMPEX, and Polar. They overpredict the degradation that spacecraft will experience from exposure to the geospace radiation environment, which results in the costly overdesign of many satellites in some orbits. A further difficulty with the standard radiation belt models in use today is that they do not capture the great variability on a variety of time scales that is characteristic of the near-Earth radiation environment. There is thus tremendous interest among spacecraft designers and mission planners in next-generation radiation belt models and in the measurements that the two Living with a Star Radiation Belt Storm Probes will provide. (Image courtesy of David Chenette, Lockheed Martin Space Systems Company.)

UNDERSTANDING THE SOCIETAL AND ECONOMIC IMPACTS OF SEVERE SPACE WEATHER

Today's society depends heavily on a variety of technologies that are susceptible to severe space weather. Strong auroral currents can disrupt and damage modern electric power grids and may contribute to the corrosion of oil and gas pipelines. Magnetic storm-driven ionospheric density disturbances interfere with HF radio communications and navigation signals from Global Positioning System satellites, while polar cap absorption events can degrade—and, during severe events, completely black out—HF communications along transpolar aviation routes, requiring aircraft flying these routes to be diverted to lower latitudes. Exposure of spacecraft to energetic particles during solar energetic particle events and radiation belt enhancements can cause temporary operational anomalies, damage critical electronics, degrade solar arrays, and blind optical systems such as imagers and star trackers.

The adverse effects of severe space weather on modern technology are well known and well documented, and risk-mitigation procedures and technologies have been developed. The physical processes underlying space weather are also generally well understood, although improvements in our ability to model the space environment and to forecast severe space weather events are needed. Less well documented and understood, however, are the potential economic and societal impacts of the disruption of critical technological systems by severe space weather—and, in particular, by rare extreme events such as the Carrington Event of 1859.

Defining and quantifying these impacts presents a number of questions and challenges with respect to the gathering of the necessary data and the methodology for assessing the risks of severe space weather disturbances as low-frequency/high-consequence events. Multiple variables must be taken into account, including the magnitude, duration, and timing of the event; the nature, severity, and extent of the collateral effects cascading through a society characterized by strong dependencies and interdependencies; the robustness and resilience of the affected infrastructures; the perception of risk on the part of policy makers and stakeholders; the risk management strategies and policies that the public and private sectors have in place; and the capability of the responsible federal, state, and local government agencies to respond to the effects of an extreme space weather event. A quantitative and comprehensive assessment of the societal and economic impacts of severe space weather will be a truly daunting task.

REFERENCES

1. Lakhina, G.S., et al., Research on historical records of geomagnetic storms, *Proceedings of the International Astronomical Union*, 2004, pp. 3-15, doi:10.1017/S1743921305000074.

2. Bedingfield, K.L., R.D. Leach, and M.B. Alexander, *Spacecraft System Failures and Anomalies Attributed to the Natural Space Environment*, NASA Reference Publication 1390, August 2006, pp. 1 and 5.

3. Silverman, S.M., and E.W. Cliver, Low-latitude auroras: The magnetic storm of 14-15 May 1921, *Journal of Atmospheric and Solar-Terrestrial Physics* 63, 523-535, 2001.

4. Walker, C.V., On magnetic storms and currents, *Philosophical Transactions of the Royal Society* 151, 89-131, 1861. The quote is from p. 95: "The fact appears to have been that the disturbance was of such magnitude and of so long continuance, and this at the busy season when the telegraph is more than usually required, that our clerks were at their wits' end to clear off the telegrams (which accumulated in their hands) by other less affected but less direct routes." Walker was the telegraph engineer for the South Eastern Railway.

5. Odenwald, S., *The 23rd Cycle: Learning to Live with a Stormy Star*, Columbia University Press, New York, 2001.

6. U.S.-Canada Power System Outage Task Force, "Final Report on the August 14, 2003 Blackout in the United States and Canada: Causes and Recommendations," April 2004. See *news.findlaw.com/hdocs/docs/ferc/uscanrpt40504ch1-3.pdf*.

7. The Carrington Event was the topic of a special issue of *Advances in Space Research* published in 2006. Unless otherwise noted, the information in this account is drawn largely from that collection of papers. The following articles were particularly helpful: Green, J.L., et al., Eyewitness reports of the great auroral storm of 1859, *Adv. Space Res.* 38, 145-153; Green, J.L., and S. Boardsen, Duration and extent of the great auroral storm of 1859, *Adv. Space Res.* 38, 130-135; Cliver, E.W., The 1859 space weather event: Then and now, *Adv. Space Res.* 38, 119-129; Shea, M.A., et al., Solar proton events for 450 years: The Carrington Event in perspective, *Adv. Space Res.* 38, 232-238; and Shea, M.A., and D.F. Smart, Compendium of the eight articles on the "Carrington Event" attributed to or written by Elias Loomis in the *American Journal of Science*, 1859-1861, *Adv. Space Res.* 38, 313-385. Elias Loomis (1811-1889) was a professor of natural philosophy at Yale University with a particular interest in meteorology. Loomis collected reports

of the aurora and magnetic disturbances observed during the 1859 storms and published them in eight installments in the *Am. J. Sci.* Henry Perkins' report is contained in Loomis' third article, which appears on pp. 322-333 of the *ASR* compendium; the description of the red aurora seen over Havana is from a report published in the first installment; it appears on p. 326 of the *ASR* compendium. The *Rocky Mountain News* account of the aurora over Colorado is quoted by Green et al., p. 149.

    8. *New York Times*, August 30, 1859.

    9. Standage, T., *The Victorian Internet: The Remarkable Story of the Telegraph and the Nineteenth Century's On-Line Pioneers,* Walker & Company, New York, 2007.

    10. The *Philadelphia Evening Bulletin* is quoted in the *New York Times* of August 30, 1859. Sparking started fires in some telegraph offices, and one operator, Frederick Royce of Washington, D.C., received "a very severe electric shock, which stunned me for a moment." A witness saw "a spark of fire jump from [Royce's] forehead to the sounder." Royce's account of his experience was reported in the *New York Times* of September 5, 1859, and reprinted by Loomis (note 7) and G.B. Prescott (note 11).

    11. Prescott, G.B, *History, Theory, and Practice of the Electric Telegraph,* Ticknor and Fields, Boston, 1860, p. 320.

    12. Carrington, R.C., Description of a singular appearance seen in the Sun on September 1, 1859, Monthly Notices of the Royal Astronomical Society 20, 13-14, 1860. Quoted in J. Bartels, Solar eruptions and their ionospheric effects—A classical observation and its new interpretation, *Journal of Terrestrial Magnetism and Atmospheric Electricity* 42, 235-239, 1937.

    13. Cliver, E.W., and L. Svalgaard, The 1859 solar-terrestrial disturbance and the current limits of extreme space weather activity, *Solar Physics* 224, 407-422, 2004. Cliver and Svalgaard rank the Carrington Event against other severe storms in terms of sudden ionospheric disturbance, solar energetic particle fluence, coronal mass ejection transit time, storm intensity, and equatorward extent of the aurora. They conclude (p. 407), "While the 1859 event has close rivals or superiors in each of the above categories of space weather activity, it is the only documented event of the last ~150 years at or near the top of all the lists."